3D Printing Designs: Design an SD Card Holder

Learn how to design 3D-printed objects that work in the real world

Joe Larson

BIRMINGHAM - MUMBAI

3D Printing Designs: Design an SD Card Holder

First published: April 2016

Production reference: 1260416

Published by Packt Publishing Ltd.

Livery Place
35 Livery Street
Birmingham B3 2PB, UK.

ISBN 978-1-78588-573-0

www.packtpub.com

Credits

Author
Joe Larson

Reviewer
Marcus Ritland

Commissioning Editor
Edward Gordon

Acquisition Editor
Vinay Argekar

Content Development Editor
Shweta Pant

Technical Editor
Vishal Mewada

Copy Editor
Madhusudan Uchil

Project Coordinator
Kinjal Bari

Proofreader
Safis Editing

Indexer
Priya Sane

Graphics
Kirk D'Penha

Production Coordinator
Shantanu N. Zagade

Cover Work
Shantanu N. Zagade

About the Author

Joe Larson is one part artist, one part mathematician, one part teacher, and one part technologist. It all started in his youth on a Commodore 64 doing BASIC programming and low resolution digital art. As technology progressed, so did Joe's dabbling, eventually taking him to 3D modeling while in high school and college, and he momentarily pursued a degree in Computer Animation. He abandoned the track for the much more sensible goal of becoming a math teacher, which he accomplished when he taught 7th grade math in Colorado. He now works as an application programmer.

When Joe first heard about 3D printing, it took root in his mind and he went back to dust off his 3D modeling skills. In 2012, he won a Makerbot Replicator 3D printer in the Tinkercad/Makerbot Chess challenge with a chess set that assembles into a robot. Since then, his designs on Thingiverse have been featured on Thingiverse, Gizmodo, Shapeways, Makezine, and other places. He currently maintains the blog `http://joesmakerbot.blogspot.in/`, documenting his adventures.

About the Reviewer

Marcus Ritland is a designer and 3D printing consultant at his small business, Denali 3D Design. Since 2008, he has provided 3D modeling and 3D printing services, as well as moderating the SketchUcation 3D printing forum.

He has volunteered at a local makerspace, teaching SketchUp classes and leading 3D printing meetups. As an author of 3D Printing with SketchUp, he is currently on a quest to eliminate design-for-3D printing illiteracy.

www.PacktPub.com

eBooks, discount offers, and more

Did you know that Packt offers eBook versions of every book published, with PDF and ePub files available? You can upgrade to the eBook version at www.PacktPub.com and as a print book customer, you are entitled to a discount on the eBook copy. Get in touch with us at customercare@packtpub.com for more details.

At www.PacktPub.com, you can also read a collection of free technical articles, sign up for a range of free newsletters and receive exclusive discounts and offers on Packt books and eBooks.

https://www2.packtpub.com/books/subscription/packtlib

Do you need instant solutions to your IT questions? PacktLib is Packt's online digital book library. Here, you can search, access, and read Packt's entire library of books.

Why subscribe?

- Fully searchable across every book published by Packt
- Copy and paste, print, and bookmark content
- On demand and accessible via a web browser

Table of Contents

Preface

3D printing makes virtual things real, and sometimes, those things need to match back up to an object in the real world. So, it is often very important that accurate measurements be made when planning and then be applied to the modeling of a 3D object. While most people don't consider Blender up to the task, it is actually capable of modeling objects with CAD-like precision. This book will get you up to speed on using Blender, then give you a project to follow along with in order to teach you the techniques to make objects that satisfy a measured precision.

What this book covers

Chapter 1, *3D Printing Basics*, will help you understand 3D printing basics, types of 3D printing, and how FFF printers work.

Chapter 2, *Beginning Blender*, will introduce Blender, how to set it up, and some basic and mid-level functionality. Knowing the content of this chapter will get you over Blender's famous learning curve and provide the basic knowledge and reference necessary for following along with future projects.

Chapter 3, *Measuring Basics*, mentions how it is very important that accurate measurements must be made when planning and applied to the modeling of a 3D object. In this chapter, we deal with different techniques of taking measurements: measuring with a ruler or calipers, the grid paper trace method, and 3D scanning.

Chapter 4, *An SD Card Holder Ring*, walks you through the process of making a cool 3D printed project—an SD card holder.

What you need for this book

A computer with at least a 2-GHz CPU, 2 GB of RAM, and, of course, Blender.

Who this book is for

This book is for anyone with an interest in home 3D printing and a desire to learn the basics of design and the tools to make their ideas a reality.

Conventions

In this book, you will find a number of text styles that distinguish between different kinds of information. Here are some examples of these styles and an explanation of their meaning.

Code words in text, database table names, folder names, filenames, file extensions, pathnames, dummy URLs, user input, and Twitter handles are shown as follows: Now, find Ch10 Scanned Image.jpg, and right-click to save the image.

New terms and **important words** are shown in bold. Words that you see on the screen, for example, in menus or dialog boxes, appear in the text like this: Start **Blender** and, as usual.

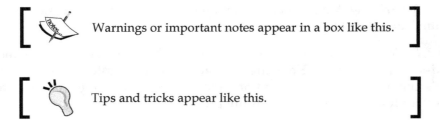

Warnings or important notes appear in a box like this.

Tips and tricks appear like this.

Reader feedback

Feedback from our readers is always welcome. Let us know what you think about this book—what you liked or disliked. Reader feedback is important for us as it helps us develop titles that you will really get the most out of.

To send us general feedback, simply e-mail feedback@packtpub.com, and mention the book's title in the subject of your message.

If there is a topic that you have expertise in and you are interested in either writing or contributing to a book, see our author guide at www.packtpub.com/authors.

Customer support

Now that you are the proud owner of a Packt book, we have a number of things to help you to get the most from your purchase.

Downloading the color images of this book

We also provide you with a PDF file that has color images of the screenshots/ diagrams used in this book. The color images will help you better understand the changes in the output. You can download this file from http://www.packtpub. com/sites/default/files/downloads/3DPrintingDesignsDesignanSDCardHold er_ColorImages.pdf.

Errata

Although we have taken every care to ensure the accuracy of our content, mistakes do happen. If you find a mistake in one of our books—maybe a mistake in the text or the code—we would be grateful if you could report this to us. By doing so, you can save other readers from frustration and help us improve subsequent versions of this book. If you find any errata, please report them by visiting http://www.packtpub. com/submit-errata, selecting your book, clicking on the **Errata Submission Form** link, and entering the details of your errata. Once your errata are verified, your submission will be accepted and the errata will be uploaded to our website or added to any list of existing errata under the Errata section of that title.

To view the previously submitted errata, go to https://www.packtpub.com/books/ content/support and enter the name of the book in the search field. The required information will appear under the **Errata** section.

Piracy

Piracy of copyrighted material on the Internet is an ongoing problem across all media. At Packt, we take the protection of our copyright and licenses very seriously. If you come across any illegal copies of our works in any form on the Internet, please provide us with the location address or website name immediately so that we can pursue a remedy.

Please contact us at copyright@packtpub.com with a link to the suspected pirated material.

We appreciate your help in protecting our authors and our ability to bring you valuable content.

Questions

If you have a problem with any aspect of this book, you can contact us at questions@packtpub.com, and we will do our best to address the problem.

1
3D Printing Basics

As cool as 3D printing is, there is a lot of hype around it, which sometimes causes confusion. Before starting to design for 3D printing, it's best to know a little bit about 3D printing technologies.

3D printing is a limitless technology in the sense that there is no end to the things it can make. Still, that doesn't mean that it can make anything without limitations. 3D printing can make things that no other manufacturing method can, but it has rules that need to be followed to ensure success. There are different types of 3D printing as well, and each type comes with its benefits and drawbacks:

In this chapter, we'll discuss:

- What is 3D printing?
- What types of 3D printing are there?
- How do FFF printers work?
- The anatomy of an FFF print.
- Supportless 3D printing and YHT.
- Wall thickness and tolerances.

What is 3D printing?

3D printing is cool. It seems as if not a day passes without another mention of 3D printing online in the news and media. Everyone is getting excited about 3D printing. But when you look deeper, it seems as if everything is being 3D printed, and anything could be. Does 3D printing something make it better? What exactly is 3D printing?

In many ways, 3D printers are just tools, the same as any that you'd find in a wood shop or garage. These tools make cool things, but not on their own, and just because something is made with, say, an electric drill press, that doesn't automatically make it better than something that isn't. It's the things that people, like you, are doing with these tools that make them cool.

I'm not saying that 3D printing isn't cool by itself. 3D printing lets you create things, test them, change their design, and try something new quickly until you get it right. It makes things of incredible complexity and, because it's additive manufacturing, generates comparatively little waste. The availability of cheaper and faster 3D printers means that there's a chance that there's a 3D printer near you.

What defines 3D printing?

There are many different types of 3D printers, but what makes them all similar is that they build solid shapes from layers of materials, starting with an empty build area and filling it with the print. This is called **additive manufacturing**, and it produces less waste than other techniques, such as starting with a base material that is cut away to make the thing.

3D printers also benefit from being computer-controlled machines, also known as **computerized numerical control (CNC)** machines, meaning they do what they do with minimal human interaction after the design work is done. They can make many identical copies of a thing one right after the other, and the design can be shared online so that others can make their own copies.

While all 3D printing shares come common features, there are several distinct types of 3D printing that vary in how they produce the print. **Fused filament fabrication (FFF)**, **powder bed**, or **light polymerization**, for example, all accomplish 3D printing in very different ways, and each with their own strengths and weaknesses. What works in powder bed 3D printing might not work with FFF 3D printing, and the part you get from light polymerization might not be suitable for the same usage as those made with the other techniques.

What to design for?

It is the best practice to always design towards the strengths and weaknesses of the medium you'll be using. The projects in this series of books will focus on designing for FFF 3D printers, because they're inexpensive and more readily available than the others, and the parts made with FFF 3D printers are suitable for a wide variety of functional uses. Also, many of the techniques for FFF design transfer to the other types of 3D printing. But because FFF 3D printers have limitations, there will be some things you need to know first.

How do FFF printers work?

There tends to be a lot of variation within the family of FFF 3D printers. Some have their mechanisms exposed to the environment so that they're easy to repair, while some are protected with fancy covers so that they look good. Some have one extruder, while some have two or more. Some have fancy interface screens, and some require you to use a computer to access even the most basic functions. Yet, for all their variations, there are many similarities that all FFF printers share which define their type. Being familiar with how FFF 3D printers work will help you guide yourself while designing for them.

For FFF 3D printers, a computer takes a 3D model and translates it into commands that the printer can follow. The printer then takes a roll of plastic filament on a spool and uses a feeder mechanism to feed it into the hot end, where the plastic filament is melted and squirted out at a controlled rate onto the print bed, where the print is built up. The extruder head and print bed are moved relative to each other in 3 dimensions, using some sort of movement system in order to create the 3D model:

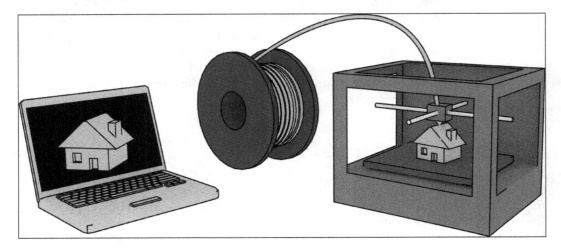

Drawing a print layer by layer like this takes, as might be expected, a little bit of time. The larger the object, the longer a print will take. FFF 3D printing isn't a fast process. But once the process is done, a new thing will have been created.

The anatomy of a print

Now that the mechanics of FFF 3D printing are clear, it's time to take a look at how a print is built. If an FFF print is stopped partway through, or observed during printing, the following can be seen:

The following are the different parts shown in the preceding image:

- **Layers**: FFF prints in layers, with each layer sitting on the one below it. Prints can be made with thicker layers so that they print faster or thinner layers so that they look better.

- **Outlines**: When starting a layer, the outline of that layer will usually be printed first. FFF prints often have two or more outlines so that the outside of the print is strong.

- **Infill**: once the outline is done, the rest of the layer is filled in. If an area of the print will not be seen from the outside when the print is done, a loose infill is used to save material and give layers above something to sit on. Top layers are filled in completely. Most FFF prints are largely hollow.

FFF design considerations

The basic limitations of FFF printers stem from the fact that most FFF 3D printers are developed by people who have very little accountability. To the people creating and manufacturing these printers, if the printer can print a thing most of the time, then that's probably good enough. In this way, FFF printers are more like garage tools than desktop machines. For those unfamiliar with FFF printers, there are some drawbacks that need to be taken into account.

Overhangs and supports

FFF 3D printers have to worry about overhang. Overhang is when a part of the design, when it prints, will not have anything between it and the build platform. To compensate for this, the 3D printer can build a lattice of support material up to the overhanging part. After the print, the support material will have to be removed. But since for most FFF 3D printers the support material is made of the same material as the object, it can rarely be removed without a trace that is sometimes difficult to clean up completely and can leave a mess on more complex prints:

Because of the troubles with supports, it's a good idea to design for supportless 3D printing.

Supportless 3D printing

Think about building a snowman or sand castle. There's a lot that can be done with the medium of sand or snow, but try to get too fancy with the design and it will fall apart. As long as every part is sitting on top of something, chances are it will hold together. You could even slope gently outwards, as long as you don't push it too far.

It's the same with 3D prints. Because it prints in layers, each layer needs to have something to lay down on. If a design is made so that a part has nothing underneath it and is dangling in the air, then the printer will still extrude some plastic to try to print the part, but with nothing to print on, the plastic will just drool from the extruder until it gets wiped off on some other part, making an ugly mess and ruining the print.

As long as you put some thought into it, you can make designs that will succeed in most cases. There are a few rules that can help, and these rules can be illustrated with the letters Y, H, and T.

Y – gentle overhangs

Think about 3D printing a capital letter Y, standing up on the build platform — something like this:

As the print gets to the part where the arms of the Y branch out, the change is gradual. It is possible to have the current layer slightly larger than the previous one, provided the overhang is gentle. Generally, a 45-degree overhang is safe. Hence, a shape like the letter Y will successfully print standing up.

However, if the overhang is too great or too abrupt, the new layer will droop, causing a print to fail. Some 3D printer owners pride themselves in pushing their overhang and have seen success with angles as steep as 80 degrees, but to be safe, keep your angles no more than 45 degrees.

H – bridging

If a part of the print has nothing above it but has something supporting it on either side, like a capital letter H standing up, then it may be able to bridge the gap when printing:

Use caution when bridging. The printer makes no special effort when making bridges; they are drawn like any other layer: outline first, then infill. As long as the outline has something to attach to on both sides, it should be fine. But if that outline is too complex or contains parts that will print in midair, it may not succeed. Being aware of bridges in the design and keeping them simple is the key to successful bridging. Even with a simple bridge, some 3D printers need a little bit more calibration to print it well.

Hence, a shape like the capital letter H will successfully print most of the time because of bridging.

T – orientation

If you were to try to print a capital letter T standing up on the build platform, you would surely run into problems:

The top arms have far too much overhang to print successfully. Of course, the solution to this is simple: when designing, flip the T over or lay it down. In fact, every letter of the alphabet will print successfully if laid on its back, but the letter T illustrates this best. Sometimes, when designing a part for 3D printing, it's good to turn it around and orient it so that it prints well. Not every print needs to be printed in the same way it's going to be used.

Wall thickness

There is a minimum size of things that a 3D printer can print. This size is determined by the size of the hole in the nozzle, called the nozzle diameter. The most common nozzle diameter is 0.4 mm; however, most printers will not print a wall with a single extrusion thickness. They require that a wall be at least as thick as two nozzle widths, which in most cases means walls need to be at least 0.8 mm. However, because of the way slicers calculate outlines, 0.8 mm isn't just a minimum wall thickness—it's a target. For instance, if the wall is 1 mm thick, it won't be able to fill in the gap between the outlines, and there will be an air pocket. And while 0.4 mm is a very common nozzle diameter, it is not the only nozzle diameter, so a 0.8 mm wall may still be too thin for some 3D printers.

For thickness, it's best to err on the side of caution. A 2 mm wall is thick enough that slicers can use one or two outlines without conflict and still have room for a little infill, no matter the nozzle diameter. This will make solid prints that will succeed in almost all cases, and 2 mm is still fairly thin, allowing for considerable detail. Unless you are designing for a specific printer or planning to share your model with others, always make your walls a minimum of 2 mm thick to be safe.

Holes in models

Models for 3D printing must be closed, that is to say, they must have no holes in them. In a classic cartoon, there was a scene where bubbles were blown, but they were not bubble shaped. They were square, squiggly, and pink-elephant shaped. But no matter their shape, they were still bubbles. If a hole developed in them, they popped. In the same way, models for 3D printing cannot have holes:

In mathematical terms, holes in models are included in a family of errors called non-manifold. Models for 3D printing must be manifold or else the slicer will have trouble telling what is supposed to be the inside and outside of the model.

In the same vein, a wall by itself, without an inside or outside, isn't printable because a 2D wall has no thickness and doesn't describe a shape that can exist in real life. 3D prints must be part of a three-dimensional shape with a thickness, as described in the previous section.

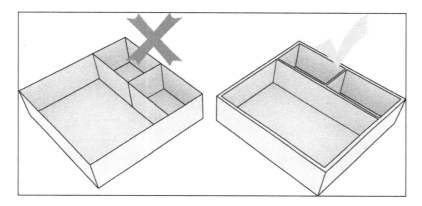

Summary

3D printing is cool and allows the creation of fantastic and detailed objects without needing much interaction with people after the design is done. But designing for 3D printing is a lot like designing for any other type of manufacturing. It helps to know a bit about the process involved and design with that process in mind.

Fused filament fabrication 3D printing, or FFF for short, is one of the oldest, most mature, and cheapest forms of 3D printing, so this series will focus on designing for it. It involves melting a plastic filament and drawing the object layer by layer, with each layer sitting on top of the one below it.

Designing for the most effective FFF printing means thinking about overhangs and supports and about the parts of the prints that don't have anything underneath them when they print. To avoid needing supports when printing, it can help to remember the letters Y, H, and T when designing, in order to remember to consider gradual overhangs, bridging, and orientation. In addition, it's important to remember that details should be, generally, about 2 mm thick.

Now that the mechanics of 3D printing and how they affect design have been covered, the next chapter will deal with the specific software that will be used in this series.

2
Beginning Blender

3D printers need 3D models to print. Those models don't just come out of nowhere. You will need imagination, a little time, and software to create 3D models for the things you want to print. Never have there been more software options for creating 3D models—professional and free options. In this series, the software of choice is called Blender.

This chapter will introduce Blender, how to set it up, and some basic and mid-level functionality. Knowing the content of this chapter will get you over Blender's infamous learning curve and provide the basic knowledge and reference material necessary to follow along with future projects. We'll cover these topics:

- Why Blender?
- Downloading and installing Blender
- The default view
- The best settings
- Object creation
- Navigating the view
- Transforming
- Controlling transformations
- Selecting
- The edit mode
- Blender to real life
- Exporting an STL

Why Blender?

With so many options for 3D modeling software, why would Blender, a software designed to make 3D animations, be the most popular choice?

The price is right

First things first: you don't have to pay for Blender. It is offered free of charge. If it works for you, you always have the option to donate, but Blender doesn't do anything to force this point. It is free now and forever.

Blender is comprehensive

While it's true that Blender is designed for animation because it covers everything from a blank canvas to a finished animation, it contains the ability to model objects, and it's one of the most robust suites of modeling tools anywhere. Learning Blender means that you may never need to learn another 3D modelling software.

It's getting better all the time

Blender is in constant development. If it doesn't have a feature you need, chances are that it may one day. Blender's developers are constantly responding to their audience.

But Blender isn't perfect

Despite Blender's advantages, it has a well-earned reputation for having a difficult-to-overcome learning curve. It's had a long and organic development cycle, which left it with a default user interface that isn't intuitive compared to most other software.

But Blender is also very configurable and, with a few simple settings, it can be made much easier for the beginner to use. The rest of this chapter will help get you past Blender's short comings so that you can start developing awesome 3D models.

Downloading and installing Blender

The first thing that needs to be done is downloading and installing Bender. Follow these steps:

1. On your PC or Mac, open a your web browser and go to http://www.blender.org.

2. Locate the **Blender** download button on the main page for the latest version of Blender and click on it:

3. Scroll down and find **Download** under **Blender** and click on it. If you're on Windows and unsure, just choose the MSI package option.

4. When the installer finishes downloading, run it.

5. Follow the prompts to install **Blender**.

6. When the installer is finished, run **Blender**. Click anywhere to close the splash screen.

Blender is now installed and ready to use.

The default view

Blender's interface is made up of many smaller windows called **panels**. There are many different panels available in Blender. Like most things in Blender, the panels are completely configurable. Panels can be added or removed as needed, and panel layouts can be saved and switched among easily. For simplicity, the default view—the way Blender is presented the first time it loads up—will be used throughout this series. It provides most of the necessary functionality:

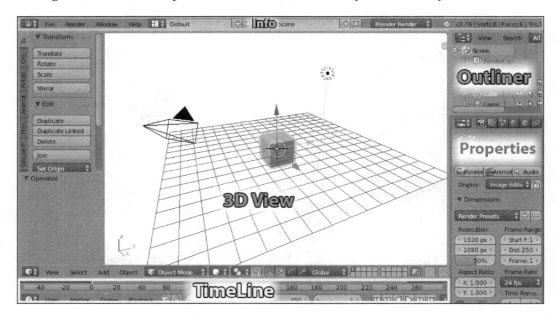

 For the most part, the screens shown in this book series will look similar to the default Blender screen. The major change will be to the background color of the 3D View, a choice made to make the illustrations more compatible with printing.

Here's an explanations of the different panels in the default view:

- **The Info panel**: Located across the top of the window, the **Info** panel has many of the menu options in most programs, such as **File**, **Window**, and **Help**. It also has **Layout** settings, **Scene** settings, and **Renderer** options, but we won't be using them much. Finally, there are specific details about the version of Blender, the current scene, and selected objects. If the **Info** panel is cut off, bringing the mouse pointer over to the panel and using the scroll wheel will bring the information into view.

- **The Outliner**: Located in the upper-right section of the window, the **Outliner** contains the list of all objects.

- **The Properties panel**: Located in the lower-right section of the window, the **Properties** panel is broken into many tabs related to the currently selected object. Which tabs and properties are available will change depending on what is selected. If the tabs are cut off, bringing the mouse pointer over them and using the scroll wheel will bring them back into view.

- **The Timeline**: This panel is largely unnecessary for the purposes of this book series, so it can be removed or ignored.

- **The 3D View**: Taking up most of the screen, the **3D View** is where most things will be happening, and it consists of many smaller parts itself.

Each of the panels has its own keyboard shortcuts. In order to use them, the mouse pointer must be over the panel. If a keyboard shortcut isn't working, it's probably because the mouse pointer isn't in the right place.

The 3D View

Because it's so complex and important, the **3D View** will be given some special attention.

The **3D View** is where most of the action takes place, and it has the most visual feedback of the work. The **3D View** has its own menus and panels:

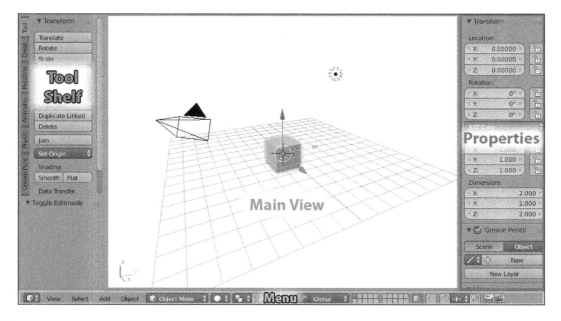

Here's what the panels do:

- **Menu**: At the bottom of the **3D View** is its **Menu**. It contains options and commands specific to the **3D View**. If the menu is cut off, bringing the mouse pointer over the panel and using the scroll wheel will bring the cut-off information into view.

- **Tool Shelf**: On the left-hand side is the **Tool Shelf**. It contains a tabulated set of buttons that can be used to do many useful things and change depending on the object selected. The bottom half of the **Tool Shelf** contains operator options. Any time an operation is performed, this area will be populated with options for that operation that can be edited until the next operation is performed. The **Tool Shelf** can be hidden or revealed by using the **3D View** menu and selecting **View | Tool Shelf** or by pressing *T* key on the keyboard.

- **Properties**: On the right-hand side is the **Properties** panel. Hidden by default, this panel contains information about the currently selected object as well as other options about the scene in general. The panel can be hidden or revealed by using the **3D View** menu and selecting **View | Properties** or by pressing *N* key on the keyboard.

- **Main View**: Of course, in the middle is the view of the current scene with all the visible objects.

The 3D cursor

In the **3D View**, there is a little red-and-white circle that starts in the middle of the 3D space. New objects will be created wherever the 3D cursor is located, and it's very easy to move the 3D cursor accidently, so it'd be good to know how to put the 3D cursor back in the middle and how to move it on purpose. The 3D cursor can be manipulated by:

1. Clicking the right mouse button in the **3D View** to move the 3D cursor wherever it is clicked in a plane relative to the view, which can be unpredictable.

2. Navigating to **View | Align View | Center Cursor and View All** or pressing *Shift + C* on the keyboard while in the **3D View** to put the 3D cursor back in the middle quickly. This is the easiest way to fix the 3D cursor if it gets moved unexpectedly.

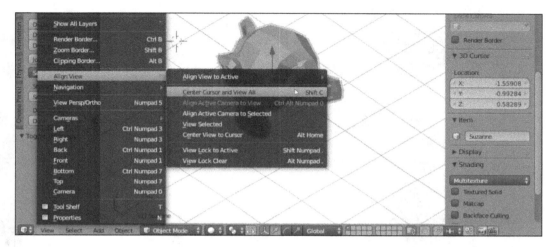

3. Typing the 3D cursor's x, y, and z coordinates in the **3D View** properties.

The best settings

Blender has a reputation for being difficult and unintuitive to use. This is true for the default settings. However, Blender is also very easily adjustable with a few settings and can be made much easier to use. How you set up your instance of Blender will depend on what your setup is like.

To access the settings, in the top menu select **File | User Preferences**. In the Blender **User Preferences** panel that comes up, select the **Input** tab button:

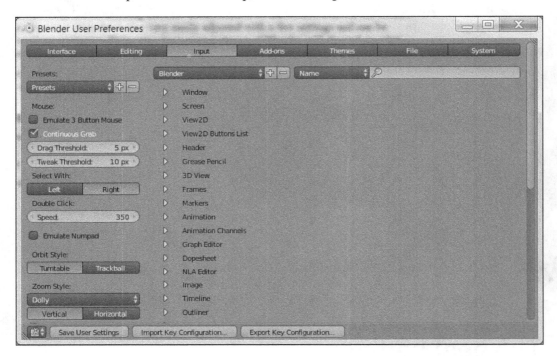

A scroll-wheel mouse and number pad

The recommended setup for Blender is to use your keyboard and a mouse with a scroll wheel. In this case, there is only one setting that is recommended to be changed from the default:

1. Click on the button that says **Left** under the words **Select With**.

2. Click on the button that says **Save User Settings**:

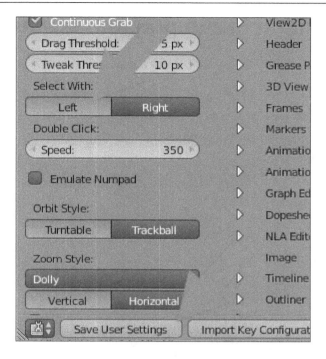

Changing this one setting will make Blender much more intuitive to use.

Because the default is for the right mouse button to be the select button, and some may prefer to keep this default, this book will refer to whatever option is chosen as the select mouse button, and the other button as the not-select mouse button. While still a bit confusing, it will help users who miss this section.

With a scroll-wheel mouse, the scroll wheel can be depressed for a middle mouse button click. Blender uses this middle mouse button to manipulate the view.

If you have a Numpad but no scroll-wheel mouse, it is still recommended to use these settings because the interface will be much more intuitive, although you will be sacrificing some functionality that the middle mouse click offers. The choice is yours whether it's worth having a more intuitive interface.

A laptop with a touch pad and no number pad

On a laptop with a touchpad (with no middle click) and no number pad, both very important for navigating the view, select the following settings:

- **Emulate 3 Button Mouse**
- **Emulate Numpad**
- Click on the button that says **Save User Settings**

With this setup, most of Blender's functionality is available to laptop users, although these settings are less intuitive. With these settings, you will need to use the right mouse button to select objects, *Ctrl* + right mouse button as a middle mouse button to change the view, and the number keys across the top will perform the function of the number pad on a regular keyboard.

With Blender set up, it's time to start learning to use it. Close the **User Preferences** window.

 Blender users tend to use keyboard shortcuts for almost everything. It's recommended that you use Blender with one hand on the mouse and the other on the keyboard. Learning and using keyboard shortcuts will speed up your process and, with just a little practice, will become second nature. Going forward, all methods of accessing commands will be taught. Try to practice using the keyboard shortcuts.

Object creation

Most tutorials for Blender start with navigating the **3D View**. But in Blender, the default scene is kind of boring for this, just a cube that looks the same from every angle. Instead, let's make something more interesting to look at.

First, let's look at how to clear the default scene:

1. Select all the objects in the scene by going to the **3D View** menu and choosing **Select | (De)select** all twice or pressing the *A* key twice. Everything in the scene should have an orange line around it. If not, do it again.

2. Delete everything by going to the **3D View** menu and choosing **Object | Delete** or pressing the *X* key.

Your **3D View** should now have nothing but the grid, which can be thought of as the floor of the scene.

To create an object, go to **3D View** menu, choose **Add** or press *Shift + A*.

In the menu that pops up are all the basic objects that can be inserted into the scene. Blender offers many basic shapes that can serve as a starting point for anything you want to create. Sometimes, a basic object is all you need, and sometimes, the basic object needs to be modified.

In this case, add a **Monkey** object to the scene:

1. In the **3D View** menu, choose **Add** or press *Shift + A*.
2. Select **Mesh | Monkey** in the menu that appears.

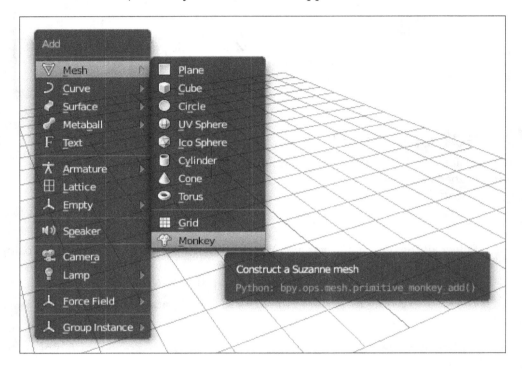

The **Monkey** object is good when learning to navigate the view because it's easy to see which way it's facing, no matter how the view changes. So, with a monkey object in the scene, it's time to learn how to change your perspective.

Navigating the view

Because Blender is all about working in 3D, but computer screens are flat, it is important to know how to change how you're looking at something in Blender.

 All of the following commands can also be found in the **3D View** menu under **View | View Navigation**; however, since adjusting the view happens so frequently, it is recommended to learn the mouse and keyboard shortcuts instead of navigating menus to do this.

Rotating the view

In Blender, you can change the angle of the view by:

- Clicking and holding the middle mouse button and moving the mouse pointer.
- Pressing *2* or *8* on the number pad to rotate the view up and down.
- Pressing *4* or *6* on the number pad to rotate the view left and right.

The point at which the view is rotating can change. Follow these steps to center the view rotation on a specific object:

1. Select the object.
2. Press the . (period) key on the number pad.

The selected object will fill the view and all view rotations will now center on that object:

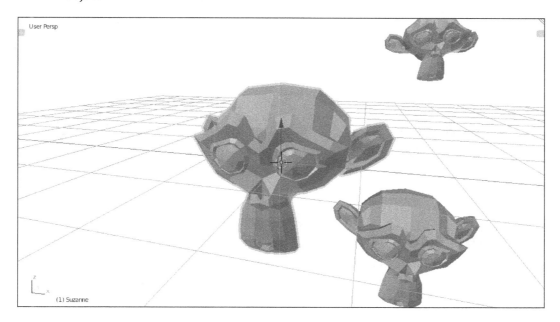

Jumping to rotation

Blender has six set views that can be quickly jumped to at any time:

- Press *1* on the number pad to jump to the front view
- Press *3* on the number pad to jump to the right view
- Press *7* on the number pad to jump to the top view
- Press *Ctrl + 7* or *9* on the number pad to jump to the bottom view
- Press *Ctrl + 1* on the number pad to jump to the back view
- Press *Ctrl + 3* to jump to the left view

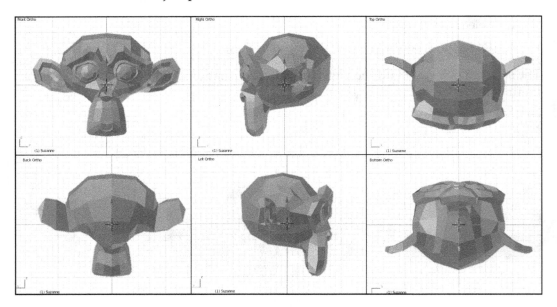

Panning the view

Panning the view means moving without changing the rotation, like moving your head while it's still pointed in the same direction. To pan the view, follow these steps:

- Press *Ctrl + 4* on the number pad to pan left
- Press *Ctrl + 6* on the number pad to pan right
- Press *Ctrl + 8* on the number pad to pan up
- Press *Ctrl + 2* on the number pad to pan down
- Hold *Shift* while clicking and holding the middle mouse button and move the mouse

Panning the view is one of the functions that changes the center of rotation. Remember: you can reset the center of rotation by selecting an object and pressing . (period) on the number pad.

Zooming the view

Zooming the view is moving closer to or farther from the object. To zoom the view, do one of the following:

1. Turn the scroll wheel.
2. Press Ctrl + middle mouse button and move the mouse.
3. Press + or – on the number pad.

Orthographic versus perspective view

To toggle between orthographic and perspective views, press 5 on the number pad.

The words **Persp** or **Ortho** can be seen in the upper right-hand corner of the **3D View**, indicating which view is being used:

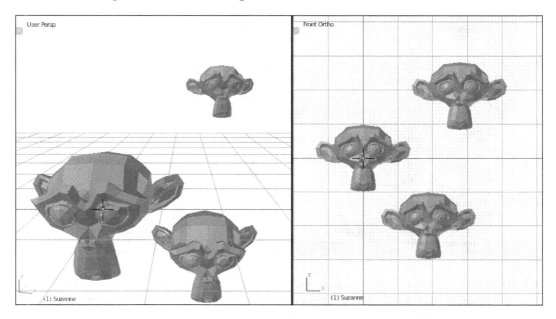

There are two ways of looking at the **3D View** on a computer. Perspective is the default, and more closely resembles how things will look when viewed through a camera or our eyes: closer objects look bigger; farther objects look smaller. Orthographic makes everything the same size, no matter how close or far it is. This makes it easier to compare objects and determine their location relative to one another.

Because a lot of 3D modelling requires precision, orthographic is generally preferred; however, it can be confusing when two objects overlap exactly. For this reason, it's good to adjust the view frequently and be sure that the action you're preforming is the action you think you're performing.

Wireframe and solid view

With 3D modelling tools, it helps to be Superman. Being able to see through objects can help when selecting and modelling things. You can switch between the **Wireframe** and **Solid** views using one of these methods:

1. Press *Z* to toggle between **Solid** and **Wireframe** view.
2. On the **3D View** menu, click on the **Method to Display** popup and choose **Solid** or **Wireframe**.

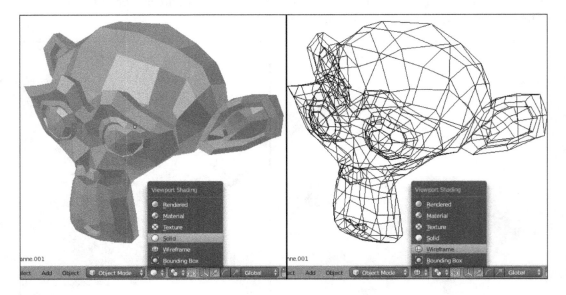

Wireframe mode is very powerful, especially in **Edit** mode, but it can be very confusing, especially as objects get more complex. **Wireframe** mode allows objects behind the objects in the foreground to be selected without adjusting the view. Get used to switching between **Solid** and **Wireframe** mode frequently.

Transforming the object

In Blender, transforming an object changes the size, direction, or location of a thing without changing its shape. There are three basic transformation commands that are used frequently: `Grab` and `Move`, `Rotate`, and `Scale`.

To transform an object, be sure the object is selected, and then, from the **3D View** menu:

- Select **Object | Transform | Grab/Move** or press *G* on the keyboard to move the object

- Select **Object | Transform | Scale** or press *S* on the keyboard to scale the object

- Select **Object | Transform | Rotate** or press *R* on the keyboard to rotate the object

Then, move the mouse or use the arrow keys to transform the object. When the transformation has been accomplished, press *Enter* or the select mouse button to end the operation, or the transformation can be cancelled by pressing the not-select mouse button or the *Esc* key. The following diagram shows these object transformations:

Transformations can also be undone after they're completed by pressing *Ctrl + Z*.

 Transform the **Monkey** added to the scene by moving, rotating, and scaling it. Get a feel for how these commands work. Then, try to move the **Monkey** to a specific place in the scene. When you think you have it, change your view to see whether it's really where you think it is.

Controlling transformations

By default, operators in Blender operate on a 2D plane tangential to the view. This is a fancy way of saying that without any additional controls, it can be hard to predict how a movement or rotation operation will work. For instance, moving something in a random view can include moving up and down more than expected. This effect won't be clear until the view is changed and the transformation is inspected from a different angle, as shown here:

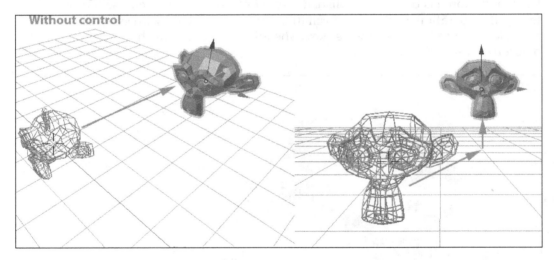

If you can't predict how operations will work, it can be hard to make the things you want. So it is very important to be able to control transformations.

There are two main ways of controlling operators: **controlling the view** and **axis locking**.

Controlling the view

The first way to control the operation is by controlling the view. By default, operators depend on the view, so by controlling the view, you can control the action. For instance, an object, when added to a scene, is exactly halfway through the grid plane of the world by default. If you want to make sure the object stays halfway through the grid plane when moving, follow these steps:

1. Select a newly added object.
2. Switch to **Top Ortho** view by pressing 7 on the number pad.
3. Grab and move (G) the object around.

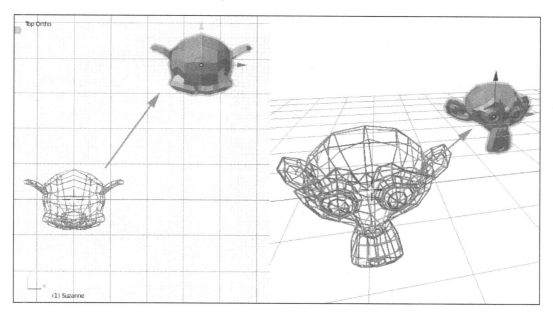

Adjust the view to see how exactly how the object moved. When moved while in the top view, the object stays on the grid plane. This is because from the top view, only forward, backward, and side-to-side motion is possible.

In the same way, moving objects while in the side views will only move them forward and back and up and down, and while in the front or back view, they will only move side to side and up and down.

Likewise, rotating depends on the view. When viewing from the top, rotation will be like, a top spinning around its middle. If viewed from the side when doing rotation, it will flip over its middle, and if viewed from the front, it will roll side to side around its middle.

Controlling the view is a quick and easy way of controlling movement and rotation, but always be sure to adjust the view to make sure things are happening the way you expect. However, to control scaling, as well as to control movement and rotation independent of the view, there is another way.

Axis locking

Blender has special commands for changing the behavior of operations. These commands are available while performing transformations.

The first thing to understand is about the three axes that we'll be locking along. 3D refers to the three dimensions, or unique directions, that things happen in. The dimensions are called by the letters x, y, and z. There is an illustration in the corner of the 3D View that shows which way is which, and is always pointing in the right direction:

In Blender, and generally in 3D printing, x is side-to-side motion, y is back-and-forth motion, and z is up and down. In the previous screenshot, the x axis is red, y is green, and z is blue.

When performing a transformation, you can lock the transformation to the axis you want by starting the transformation as mentioned previously, and while moving the mouse or using the keyboard, doing the following:

- Pressing *X*, *Y*, or *Z* on your keyboard to lock the transformation to the desired axis
- Pressing *Ctrl + X*, *Ctrl + Y*, or *Ctrl + Z* to lock the transformation to all but the chosen axis

- Holding the middle mouse button and moving the mouse to choose an axis to lock the transformation to

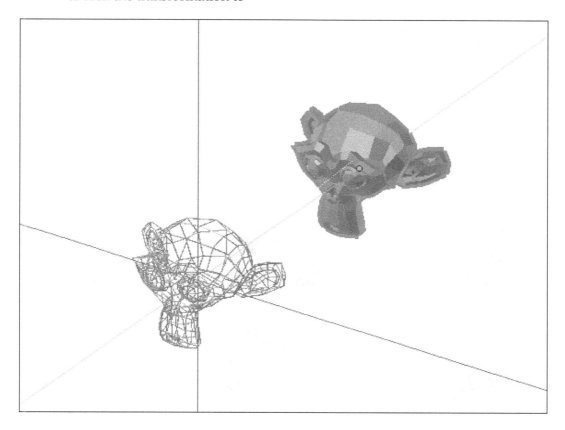

Try out axis locking with movement, rotation, and scaling, and adjust the view to see what effect it has. Notice that with axis locking, the scale function has some additional abilities that aren't possible just by controlling the view. It actually only scales in the chosen axis. This can be very powerful while adjusting the shape of objects.

Precise transformation

Finally, under the category of controlling transformations, during transformation operations, the transformation can be precisely controlled by typing in a number that will relate to the operation being performed or edited afterwards in the operation properties in the tool box, on the left side of the **3D View**. In other words, to move something exactly two units up, for example, follow these steps:

1. Select the object.
2. Start the movement transformation (*G*).
3. Type *Z* to lock the movement to the z axis.
4. Type *2* on the keyboard.
5. Press *Enter* or click the select mouse button to finish the operation.

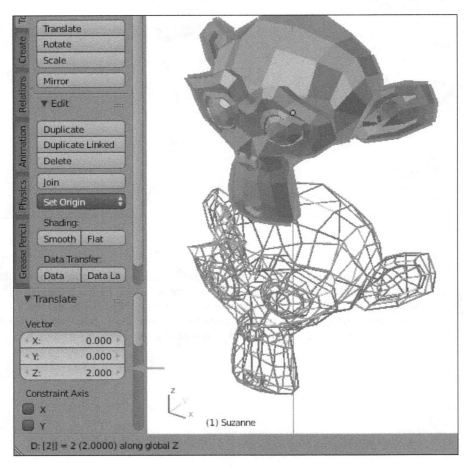

Typing commands has different effects for different commands, as follows:

- When moving, typed commands are used to state the number of units an object will move along the selected axis. For example, 2 along the z moves it two units up, and -2 along the z moves it two units down.

- When scaling, typed commands specify the scale factor. 1 means no change, 2 means twice as big, and 0.5 means half size.

- When rotating, typed commands specify clockwise degrees. For instance, 180 turns it around backwards.

Typed commands can include negative numbers and decimals and can be edited with the *Backspace* key. While performing a transformation, the menu for the **3D View** changes to a description of the transformation, including the typed units, which can be useful when editing your typed commands.

Origin manipulation

Objects in Blender have an origin. Origins are depicted as a dot and start out in the middle of the object. Individual object transformation commands take place relative to the object's origin. It's possible that the origin can be moved as a result of editing, which can cause unexpected results when rotating or scaling the object. The origin can also be moved on purpose to control the effect of modifiers.

The origin can be reset using the origin controls. The origin controls can be found by one of the following methods:

1. In the **3D View** menu, choose **Object | Transform** and find the origin controls.

2. In the **Tool Shelf**, find the **Set Origin** dropdown and choose the desired option.

3. On the keyboard, press *Ctrl* + *Shift* + *Alt* + *C*.

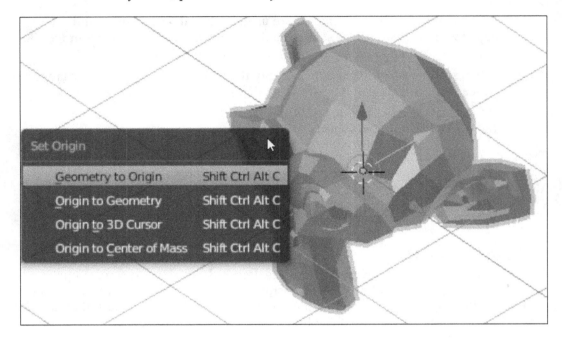

These are the commands to manipulate the origin:

- **Geometry to Origin**: Move the object so that its middle is now wherever the origin was located.

- **Origin to geometry**: Move the origin to the middle of the object. This is the most common option chosen.

- **Origin to 3D Cursor**: Relocate the origin to where the 3D cursor is.

- **Origin to Center of Mass**: Calculate the center of mass of the object and move the origin there.

Duplicating objects

Blender can duplicate existing objects. This is very useful and can speed up making things. To duplicate an object, follow these steps:

1. Select the object to be duplicated.

2. In the **3D** menu, choose **Object** | **Duplicate** or press *Shift* + *D*.

3. Move the mouse or use the arrow keys to place the duplicate (axis-constraining commands also work at this point).

4.   Press *Enter* or the select mouse button.

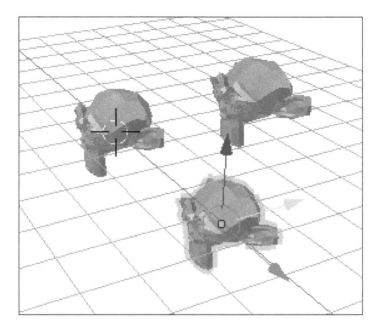

Object selection

Naturally, Blender is capable of selecting more than one object at a time. Blender has many tools to help when selecting objects. With multiple objects, the transformation commands work on all the selected objects at once. This is a powerful way of controlling transformations while keeping objects in relative positions to each other.

 To practice selection, there need to be multiple objects to select. Create a scene, add an object, and then use the duplicate (*Shift + D*) command to create many objects from the one. Spread them around the scene so that they can be selected with the following tools.

Shift select

One way to select multiple objects at once is to hold down the *Shift* key while clicking on the desired objects one at a time. To deselect an object, it must first be made active. The active object is highlighted in a different color. Then it can be clicked again while holding the *Shift* key, and it will be removed from the selection.

Border select

Another way to easily select multiple objects in a scene is to use the border select command and draw a box around the objects you want to select. To border select, choose **Select | Border Select from the 3D View** menu or press *B* on your keyboard. Then, click and hold down the select mouse button, move the mouse pointer, and release the mouse button:

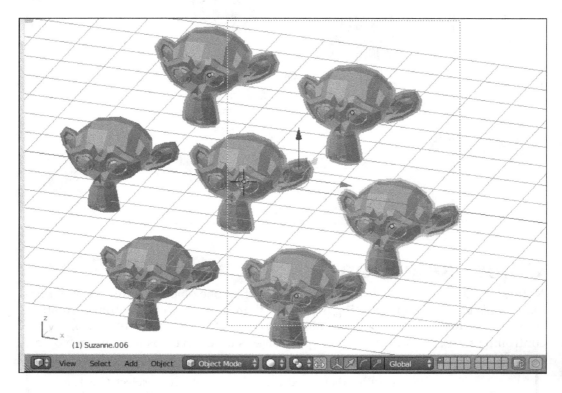

If even a small part of an object is included in the box, it will be added to the selection.

To deselect objects with border select, use the middle mouse button when drawing the box.

Circle select

Circle select is another powerful tool for selecting multiple objects. To circle select, choose **Select | Circle Select from the 3D View** menu or press *C* on the keyboard. A circle will appear around the mouse pointer. Use the scroll wheel or + and – on the Numpad to increase or decrease the size of the circle. Then, click or click and hold the select mouse button, and everything inside the circle will be selected. Use the middle mouse button to deselect:

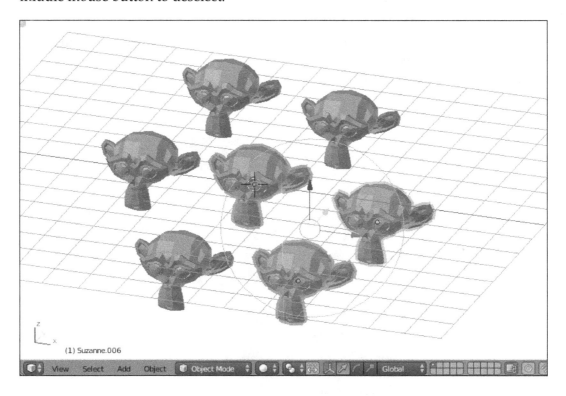

Circle select only adds or removes an object from the selection if its center or origin is inside the circle when selecting.

The Edit mode

In Blender, the **Edit** mode allows more access to the shape of a single object so that it can be manipulated in order to change its shape. To enter **Edit** mode, use this method:

1. Select an object.
2. In the **3D View** menu, locate the mode pop-up menu and select **Edit Mode** or press *Tab* on your keyboard.

In **Edit** mode, the **3D View** menu, **Tool Shelf**, and **Properties** all change, adding new functionality only available in **Edit** mode:

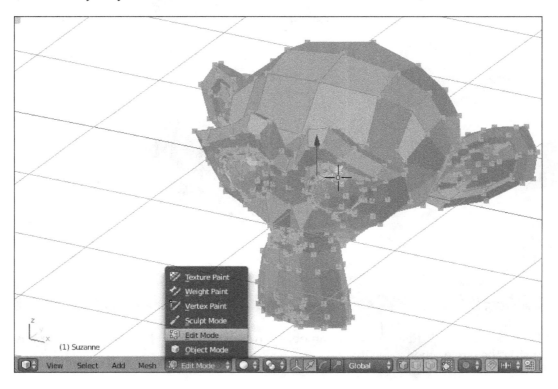

Parts of objects

In **Edit** mode, objects are broken down into three parts:

- **Vertices**: Points in three-dimensional space. Vertices don't have any shape by themselves
- **Lines**: Two points are connected with a straight line between them
- **Faces**: Three or more lines can be connected to make a face

There are many ways to think about vertices, lines, and faces. For instance, if making a kite, the vertices are the joints, the lines are the sticks, and the bits of paper are the faces. If the location of the vertices is moved, the shape of the kite will change. It's the same with a 3D object. Editing by vertices, lines, or faces will affect the rest.

The **Edit** mode is by default in vertex select mode, meaning any selection or transformation is applied to vertices, but it's easy to switch by locating the vertex, line, or face select buttons in the **3D View** menu or by pressing *Ctrl + Tab* and selecting the desired option from the menu that pops up:

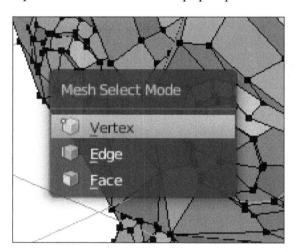

Try out **Select** mode. Select groups of vertices, edges, and faces, using circle and border select. Try moving, rotating, and scaling them and see how it effects the shape of the object. Try simple basic shapes such as cubes and circles. See how **Wireframe** mode affects selection. Notice how the other objects in the scene can't be accessed.

A lot of what happens in the projects will happen in **Edit** mode. But the individual projects will better teach you how to use it.

Incremental saving

It is always a good idea to save your work frequently. To save your work in Blender, choose **File | Save** from the info panel or press *Ctrl + S*. Then, navigate to a chosen folder or directory, give the file a name, and click on the **Save As Blender File** button or press the *Enter* key.

It is recommended that each Blender project gets its own folder and that all projects be saved in a location that will be easy to find later.

It is also a good idea, while learning especially, to give each version of the file you save a slightly different name. This way, there is a history of the work done and it's easy to go back in the case of a mistake that isn't discovered immediately. This is called incremental saving and is simple to do in Blender. Simply choose **File | Save As** or press *Ctrl + Shift + S* to get to the save menu. If the project has been saved previously, it should already have a name. Next to the filename, there are plus and minus buttons:

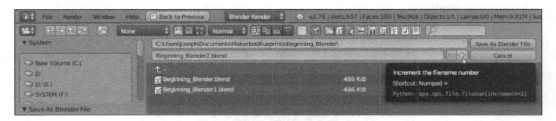

By clicking on those buttons, the filename will have a number attached to it, and that number will be increased every time the plus button is clicked. Then, click on **Save as Blender File** or press the *Enter* key to save the file with a new filename.

Blender to real life

By default, Blender units don't make any attempt to relate to real-life measurements. However, after exporting a mesh, the slicing software will interpret the Blender units as millimeters, generally. So, it is good to think of Blender units as millimeters. This means that default objects in Blender are 2 mm across when they're added, which is fairly small.

Remember that one blender unit will be 1 mm when the file is used to 3D print an object.

Exporting an STL

Before they can be used in a 3D printer, 3D models created in Blender have to be changed to a file that the 3D printer can use. Blender's default file format isn't readable by 3D printers and can sometimes contain additional information that the 3D printers don't need.

Most 3D printers use a file format called **STereoLithography** (**STL**), which contains just the final shape of the object. To export a finished model to an STL for 3D printing, follow these steps:

1. Select the model or models to be exported.

2. From the **Info** panel, choose **File | Export | Stl (.stl)**.

3. Navigate to a chosen folder or directory.

4. Give the file a name.

5. Click on the **Export STL** button or press the *Enter* key:

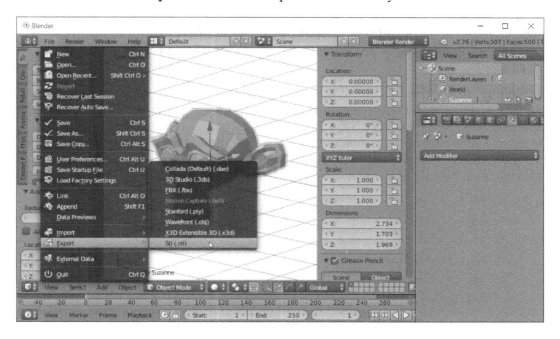

Now, the STL file will be available to send to a 3D printer or printer service.

If multiple objects are selected, the exported STL will have the objects in the same relative orientation to each other; therefore, it's important that they don't overlap and are printable as oriented. It's often preferable to save separate objects in separate files.

Summary

A recent poll of online social sites about 3D printing showed Blender was the most popular choice for creating models for 3D printing. The reason is obvious, taking into account Blender's vast functionality. However, it could just as well be that more people were talking about Blender because of its challenging learning curve.

Blender is capable of creating simple primitive shapes, viewing them from any angle, transforming them with precision, and manipulating their individual vertices, edges, and faces in the powerful editing mode. The model can then be exported to a file, ready to be 3D printed.

Blender has many functions not even covered in this chapter, such as sculpting, skeletal manipulation, and how to use individual modifiers to achieve specific results. As these functions become important for individual projects, they will be covered.

Hopefully, this chapter served to introduce how powerful and comprehensive Blender is. However, Blender's comprehensive nature comes at the cost of being complex, which can be overwhelming. But don't worry. Keep this chapter at hand to act as a reference and a crutch in future projects until Blender's functions become second nature. It usually only takes one project before Blender becomes second nature. Once Blender is familiar, no other 3D modelling software will be necessary.

3
Measuring Basics

3D printing makes virtual things real. Sometimes those things need to match back up with an object in the real world after they are made, such as a lid for a can or a joint between two poles. Sometimes you're printing a replacement for a broken part. So it is often very important that, when planning, accurate measurements are made and applied to the modeling of a 3D object.

Some measuring tools you may already have around your house. Some may require you to buy special tools you don't have but should probably consider getting.

In this chapter, we will cover:

- Measuring with a ruler
- Measuring with calipers
- Grid paper trace method
- 3D scanning

Measuring with a ruler

Most homes have a ruler on hand that can, in many cases, provide relatively accurate measurements. For irregularly shaped objects, it may be difficult to use a ruler. But if the object is flat on at least two adjacent sides, rulers work fine in a pinch:

Simply lay the object on its flat edge, move one side to line up with the 0 mark, and measure the other side of the object. It's not the best method for measuring complex shapes, but sometimes a ruler can be enough.

Measuring with calipers

The most common technique for taking accurate measurements is the use of a tool called a **caliper**, a must-have for anyone who models for 3D printing. Calipers measure distance with a high degree of precision and can measure in three different ways: the outside diameter of an object with the outside jaws, the internal diameter with the inside jaws, or the depth with the depth probe at the far end. There are two common calipers: manual (or Vernier) calipers and digital calipers.

Manual or Vernier calipers

If budget is a concern, then perhaps a Vernier caliper is preferable, since they are generally less expensive than their digital counterparts. These calipers operate purely mechanically, but have a clever trick that allows them to be just as accurate if read properly:

Open the jaws to take a measurement and tighten them over the object to be measured. Then take a close look at the little window:

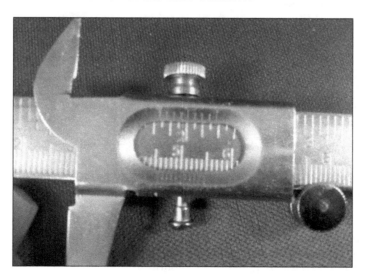

The leftmost tick is between 4.5 and 4.6, so this object is between 4.5 cm and 4.6 cm (or 45/46 mm). Then, count the tick marks on the outside until one of them lines up perfectly with one of the ticks on the inside. In this case the fifth tick mark on the outside lines up with a tick mark on the inside. This is the hundredths part of the measurement, so this object measures 4.55 cm or 45.5 mm.

Vernier calipers never need to be calibrated. They don't need batteries, so they'll always work, and they're cheaper. But they do take some additional effort to read properly and lack the "cool" factor of their digital cousins.

Digital calipers

The easiest measuring tool to use is the digital caliper. Simply turn it on, tare or zero the reading while closed, then open the jaws, put the object to be measured between them, clamp it down, and take the reading. It's fast and accurate without much effort. Some models even have a port that can transfer the measurements directly to the computer:

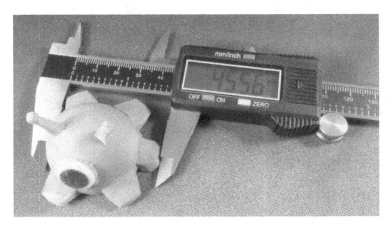

But digital calipers have the disadvantage of being more expensive and relying on batteries. They can also go off calibration, reducing their accuracy. However, for quick and easy use, a digital caliper is hard to beat.

Whatever calipers you use, familiarize yourself with all parts for measuring different things. The large jaws on the front are for measuring the outside of objects. The small jaws opposite them are for measuring the inside of objects. At the end of the ruler section is a depth sensor that can be used to measuring the depth of something.

Grid paper trace method

There is another trick to measuring complex shapes that involves a common household item. An object with a complex shape can be traced on a piece of grid paper. That grid paper trace can then be scanned or photographed, uploaded to the computer, and imported into modeling software to recreate its shape. By tracing the shape on graph paper, there is a scale reference in the modeling software.

Suitable objects

The ideal object for this method of measuring is an object with a complex and otherwise difficult to measure shape but with at least one flat side so that it's easy to lay flat on a piece of paper. If the object doesn't have a flat side, being able to make one side flat will do the trick.

For example, this piece of plastic attaches to the back of a drawer to guide it along a rail so that it moves smoothly and straight:

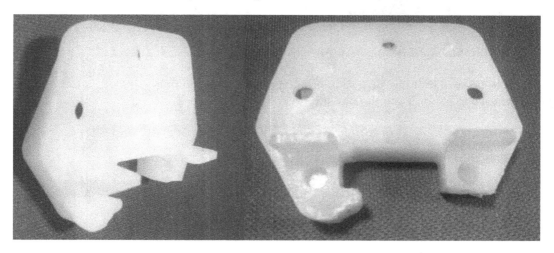

The back of this piece is almost flat, except for 2 nubs sticking out, which would be easy enough to remove. Simply note the location of parts that are removed before removing them, since they'll need to be modeled back in.

Object preparation

To begin, use a sharp blade and cut the nubs off the mostly flat surface:

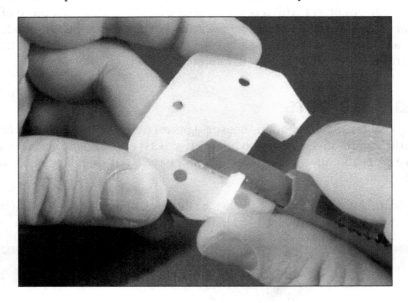

Then, to be sure that the surface is perfectly flat, use some fine-grit sandpaper on a flat surface to sand down the flat side:

Next, take some grid paper, centimeter if available, and lay the object on the paper. Line it up with the grids as best as possible, and trace the object. Keep in mind that traces are generally a little bigger than the actual objects.

As an alternative, an ink pad can be used to make an even more accurate shape transfer. Just use the flat side of the object like a stamp. Again, try to line the part up with the grid as much as possible when making the impression in order to make the later modeling process easier. Stamping on a soft surface such as a towel can help get better coverage in the stamping process:

Finally, scan the stamped part in, or if a scanner isn't available, use a digital camera — whatever is possible to get the traced-image-on-graph-paper reference into the computer. Try to keep the grid lined up and even in the image, or some editing of the image will be necessary. It is not possible to edit the image in the modeling software beyond some basic transformations, so keep it as straight horizontally and vertically as you can and avoid skewing and warping of the image as much as possible:

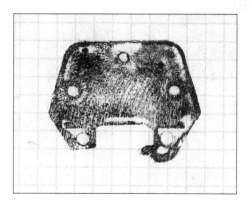

Importing the image into Blender

To follow along from this point, point your browser to `http://thingiverse.com/thing:90754`.

Now, find `Ch10 Scanned Image.jpg`, and right-click to save the image:

1. Start **Blender** and, as usual, clear the scene (*Ctrl* + *A* + *X*) and save it to start a project.

2. Give it an appropriate name in an appropriate directory, such as `Ch10 Measuring and Drawer Guide.blend`.

3. Change the view to **Top Ortho** (Numpad *7*, Numpad *5*).

4. In the **Properties** panel (*N*), locate the **Background Images** section, check the box next to it, expand it, and click on the **Add Image** button:

5. Click on the **Open** button and navigate to where the scanned image of the stamped object is stored. Then, open the image to place it in the scene:

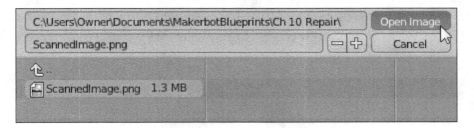

 The location and name of the scanned image may be different for you.

6. The first thing to note is that if the previous steps are followed carefully, despite the grid paper in the image being a centimeter grid, at the default zoom, the grid lines seen are in millimeters. There are 10 millimeters per centimeter, so zoom out until the major grid lines start to appear, and then zoom out some more so that the millimeter grid lines disappear:

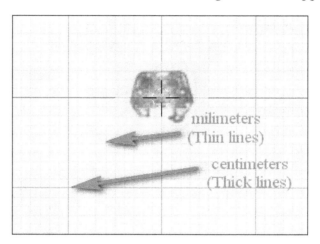

Adjust the settings in the **Property** panel for the background image until the grid on the paper lines up with the centimeter grid in **3D View**. A quick way to determine the necessary size is to count the number of grid lines in the scan, multiply it by 10, and divide by 2 (for the radius, which for some reason the size setting is). In the example, there are 12 grid lines shown, so setting the size to 60 gets the view very close to the right size.

7. Continue to adjust the **X**, **Y**, **Size**, and **Rotation** values as necessary until the grid lines in the image line up with the grid lines in the view properly:

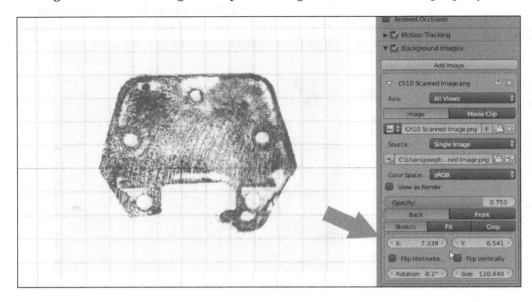

8. Since the object is to be symmetrical in this example, it would probably be a good idea to use the **Mirror** modifier, and to use it properly, it would be easier if the center line of the object were lined up with the world origin. The **X** value can now be adjusted to make it so. It no longer matters if the **X** grid lines line up with the image:

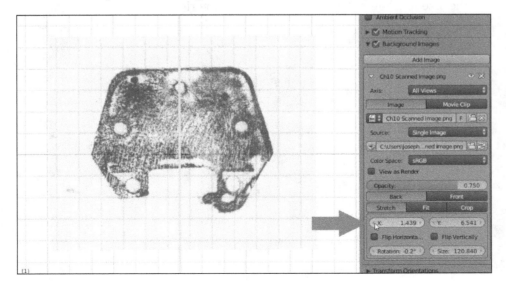

Now that the reference image is in Blender and scaled properly, the process of creating a shape in Blender around this reference image is simple and left to you.

9. Since the scale of the object was determined by lining up the real-life grid, the printed part should be approximately the same as the original piece:

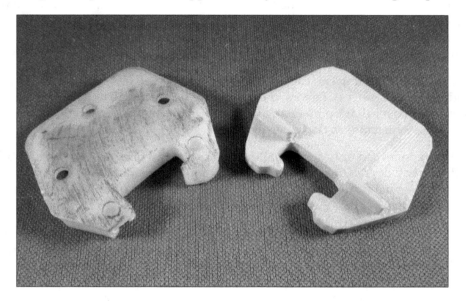

The original is still a little messy after being used as a stamp, but the new one will take its place nicely.

Increasing reference pictures

In the background image settings, the pictures can be limited to specific views: top, bottom, front, left-hand side, right-hand side, and so on. Using these settings, multiple reference images can be combined to increase the accuracy of the model. If the shape is such that it's difficult to trace accurately from the side or front, it can be hard to trust the references. In these cases, the other images can be thought of as guidelines to be followed, rather than a reference to be trusted.

3D scanning

Finally, there is a method of capturing complex shapes that can quickly and accurately reproduce the shape on the computer, but at a considerably higher cost. There are many ways to 3D-scan objects.

Photogrammetry, or building a model from photographic references, captures the details of the shape of the object, but doesn't capture the scale very well. It captures small or big objects without distinction, but the resulting model can't be used for measuring purposes very well.

Structured light scanning can capture the scale pretty well if it's calibrated, but depending on a number of factors, the geometry may lack fine details. Often, scanners may cover that up by texturing the object, but textures don't print on FFF printers, which makes the cover-up obvious.

Some 3D scanners can only scan small objects. Some can only capture larger objects, but without any degree of small detail. Some require considerable effort on the part of the user to achieve any results. And if a 3D scanner does offer high detail and measuring accuracy, chances are it comes at a cost prohibitive to home users.

3D printing has driven up demand for accurate and cheap 3D scanners. There are many options available and new ones being developed all the time. As of the time of writing this, there is no clear family of winners that can be pointed to yet. Be cautious. Many promises are being made; not as many promises are being delivered on. But keep an eye out because, one day, a reliable, accurate, and affordable 3D scanner is sure to be made.

Summary

Truth is, there's not much practical point to 3D printing if it can't be used to make real-life objects that conform to real life. Making measurements and applying those measurements to virtual objects is mandatory in 3D design. A ruler will do the trick in some cases, but more complex objects will require more robust tools. Calipers are a designer's best friend most of the time. Tracing the object onto graph paper can help transfer a complex shape into the computer, where it can be recreated. Finally, for the professional, a 3D scanner can quickly transfer a complex shape, but at considerable cost. The more options you have available, the better designer you will be.

In the next chapter, we'll use measurements to create a 3D model that will match a real-life object when we're done.

4

An SD Card Holder Ring

Many 3D printers can print directly from an SD card instead of being hooked up to a computer all the time. This presents a problem of transporting SD cards without tying up your hands. Or pockets. A ring that you can put an SD card in is just the thing.

While a ring you can put an SD card in is a slightly silly solution, it's an excellent example of modeling based on physical objects. Both a finger and an SD card will need to be measured, and the design will have to be modeled too:

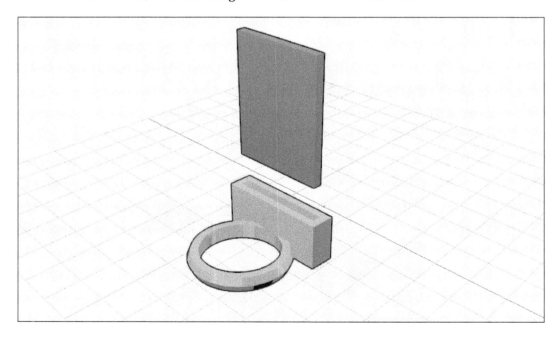

This chapter will walk you through the process of making a cool 3D-printed project, going through the following steps:

- Taking measurements
- Modeling the ring
- Testing the ring
- Adding an SD card holder
- Putting it all together
- Extra credit

The challenges of matching real-life measurements to 3D-printed objects is made difficult by plastic shrinkage and inaccuracies in 3D printers' operations. I'll comment on that when it becomes important.

Some would claim that Blender isn't capable of CAD-like precision, but this project will put those naysayers to rest.

Precise placement of new objects is important in this project. Remember that if the 3D cursor is ever moved with an accidental left mouse click, then navigate to **View | Align View | Center Cursor and View All** or press *Shift* + C to put the 3D cursor back at the origin.

Taking measurements

Before opening Blender, some measurements must be gathered. Careful measurements need to be taken of the ring finger and an SD card. A digital caliper is an excellent tool to take measurements with, as shown in the following figure:

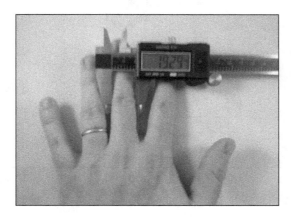

My middle finger measures 19.3 mm at the widest point. If a digital caliper is not available, another way of measuring your finger is to wrap a piece of paper around the finger, mark where it overlaps, and use a ruler to find your finger's circumference, like this:

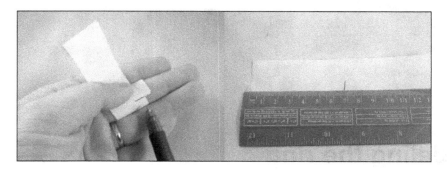

Then, consult the following table to find out the standard ring size and diameter:

Circumference (mm)	Circumference (in)	Diameter (mm)	Radius (mm)	US standard ring size
1.94	49.3	15.7	7.85	5
2.04	51.9	16.51	8.25	6
2.14	54.4	17.32	8.66	7
2.24	57	18.14	9.07	8
2.34	59.5	18.95	9.47	9
2.44	62.1	19.76	9.88	10
2.54	64.6	20.57	10.28	11
2.65	67.2	21.39	10.69	12
2.75	69.7	22.2	11.1	13
2.85	72.3	23.01	11.50	14
2.95	74.8	23.83	11.91	15
3.05	77.4	24.64	12.32	16

An adult male is generally around a size 10 mm, an adult female is around a size 8 mm, and a child is around a size 5 mm.

The next thing to measure is a standard SD card. Fortunately, SD cards are all the same: 2.2 mm x 24 mm x 32 mm:

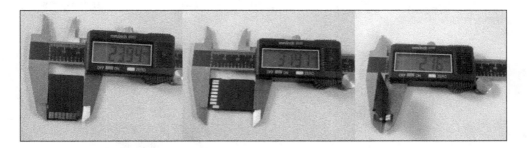

Modeling the ring

Now that all the measurements have been taken, it is time to go to Blender. Start a new scene and clear all the default elements (*A* + *A* + *X*). Then, save the scene (*Ctrl* + *S*) to create a new project in the directory of your choice. Name the project SD Card Ring.blend.

Modeling the finger

Let's start the modeling with the following steps:

1. Add (*Shift* + *A*) a Cylinder.
2. In the left-hand side bar, change the options for the cylinder. Change the **Vertices** value to 64 so the cylinder is smoother than default. Change the **Radius** value to half the measured finger diameter. Change the **Depth** value to 10:

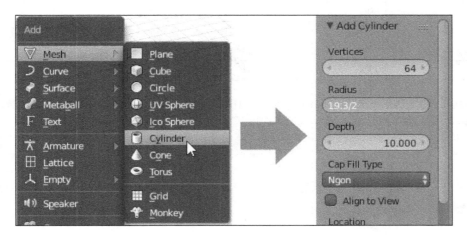

Blender can take simple equations in these boxes, so instead of dividing the diameter by half to get the radius, simply type the diameter followed by /2, and it will divide it by 2 for you. So, in the example, typing 19.3/2 resulted in a radius of 9.650.

3. In the properties tab in the right-hand side bar, in the **Object** tab (orange cube), rename this cylinder to **Finger**, since it represents the finger in the build space:

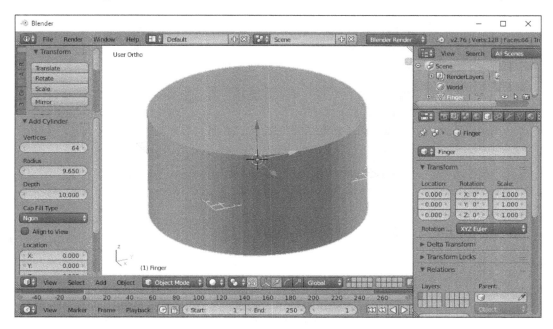

4. Now, add (*Shift* + *A*) another cylinder.

    Blender remembers the settings that the previous object was created with, and if another object is created of the same type, those settings will be applied to it. So, the new cylinder is of the same radius, depth, and number of vertices as the **Finger** cylinder, and is therefore in the same exact place until the settings are changed.

5. Add 2 to the **Radius** value of the finger and change the **Depth** value to 4:

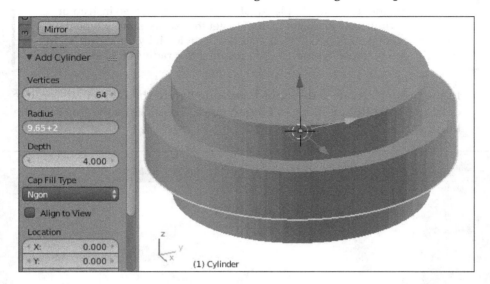

Again, adding 2 to the existing diameter is as simple as clicking on it and appending +2 to the end and then pressing *Enter*.

6. In the **Object** tab, rename this cylinder to Ring.

The **XY** plane is the logical floor of the build space and passes through the origin. Blender places a (small) **XY** plane in the view space to help with visualization. It is, of course, completely imaginary, and objects can be built through this floor. In fact, by default, new objects are placed somewhat under the **XY** plane. However, respecting the **XY** plane can provide a solid base of reference for building objects if need be, especially when multiple objects are to be placed with precision.

7. Now, jump to the front (Numpad *1*) orthographic (Numpad *5*) view. You will notice that the ring is sitting half above and half below the origin:

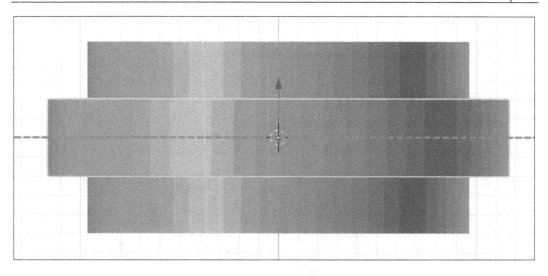

Putting the ring on the floor

By default, objects in Blender are created centered at the origin, meaning a part of them is under the **XY** plane. The goal of this section will be to place the important objects so they sit on the **XY** plane without going below it. This way, we always know where the bottom of the object is and can line everything up to this standard. So, with only the ring object selected, begin the grab (*G*) operation. Press *Z* to lock the ring's movement along the z axis. Type 2 to move the ring up half its depth. Finally, press *Enter* to end the grab operation:

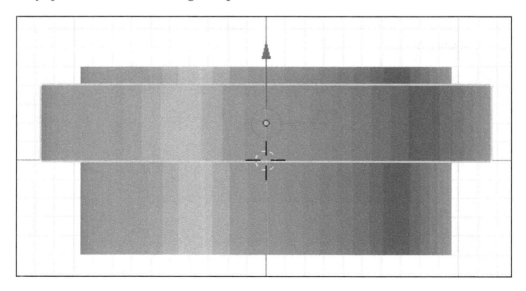

The finger cylinder is at an odd place in relation to the ring. But since the finger cylinder is just there to help us design the ring, and since it's both above and below the ring cylinder, it can just be left where it is.

Finishing the ring

Perform the following steps to complete the ring:

1. With the `Ring` object selected, add a **Boolean** modifier to it:

2. Change the the **Operation** field of the **Boolean** modifier to **Difference** and the **Object** field to **Finger**:

The **Finger** object is obscuring the effect of this operation, so it should be hidden. This may take the object out of the **3D View**, but it can still be selected in the **Outliner** panel and modifications can be made that way. Some editing remains impossible, however, until the object is unhidden.

3. Select the **Finger** object and, from the menu at the bottom of the **3D View** panel, navigate to **Object | Show/Hide | Hide Selected**, or press *H*. Another way is to click the icon that looks like an eye next to the **Finger** object in the **Outliner** panel:

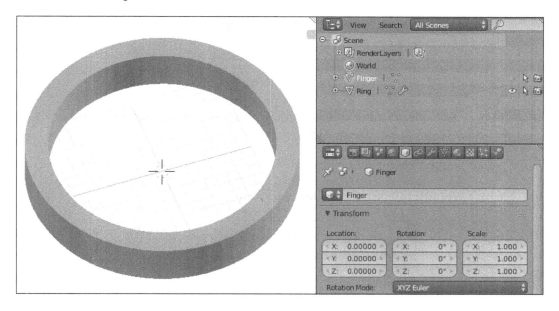

Now, the **Finger** object is hidden and we can see the effect of the **Boolean** modifier on the Ring object. The **Finger** object is still in the scene, but not in the **3D View**. You can confirm this because the object is still in the **Outliner** panel, but the eye icon beside it is inactive.

The ring is now perfectly serviceable, but there's no harm in making it a little more appealing.

4. Enter **Edit** mode (*Tab*). For now, loop cut (*Ctrl + R*) the cylinder around the middle and left-mouse click twice when selecting so that the slice does not slide up or down:

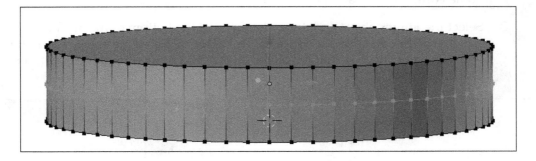

 Without applying the modifier, its effects are undone while in **Edit** mode, and the hole in the middle is gone. This is okay. When we exit **Edit** mode later, the hole in the middle will be back.

5. Scale (*S*) the newly sliced ring outwards slightly. The scaled ring of points should not extend more than 1 mm (one small grid square in the background) larger than the top and bottom disks of the ring shape.

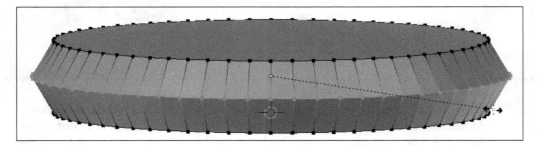

Scaling is an operation for which precise measured adjustments can be difficult to accomplish. This is one of the places where Blender falls a little short as a CAD-precise tool. Then again, this is a change made for aesthetics, so just eyeball the angle. Anything less than 45 degrees will be fine.

6. Now, exit **Edit** mode (*Tab*). Notice that the hole is back in the ring:

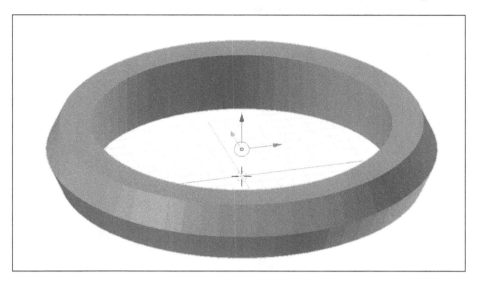

7. Remember to save (*Ctrl + S*) the project periodically.

Making a test print

If you are just learning to design without access to a 3D printer, then you can skip this step. However, since this print is so small, it is cheap, even through 3D-printing services. The ability to quickly make things for testing is one of 3D printing's greatest strengths, but it's not strictly necessary to learn how to design for them.

It's always good, especially when matching things to real-life objects, to test whether the fit is right. Some 3D printers print things smaller or larger than the file dictates. So, while the thing is small, and since 3D printers are so good at quickly testing things, test the ring you've made.

Export an **STereoLithography (STL)** of the ring through the **Export STL** option under **File | Export**. Name the exported file `Test Ring.stl`. Then, print the test ring and try it on:

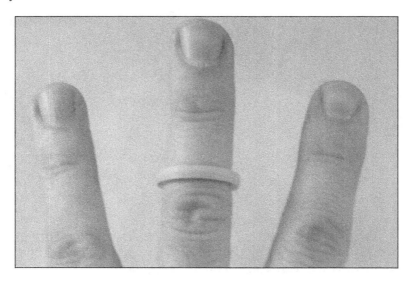

It might come as a surprise that the ring doesn't fit. If printing in ABS, the plastic may have shrunk when it cooled, as that is one of the properties of ABS. Or there may be a print setting such as filament diameter that isn't correctly set. Also, internal rings created by polygons are known to be a little small sometimes. Whatever may be the reason for the ring not fitting correctly, it's a good thing it was tested while the project was still simple.

Resizing the test ring

Back to Blender to make a new test ring. This time, instead of starting from scratch, reload the project and resize the existing ring. Take care to ensure that while resizing, the ring doesn't extend below the **XY** plane. So, to avoid this, we're going to resize the ring (and finger hole) from the **Properties** panel, like this:

1. Start by unhiding the finger. Do this by either clicking the eye icon for the **Finger** object in the **Outliner** panel or by pressing *Alt + H*:

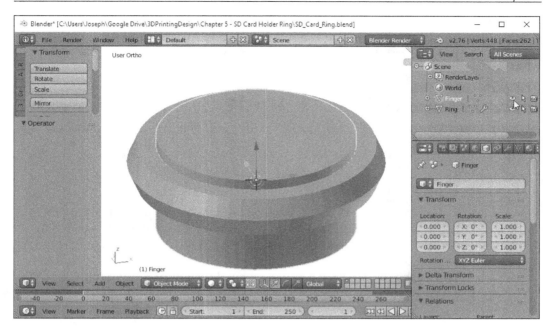

2. Select the **Finger** object and, in the **Properties** panel (*N*), locate the **Dimensions** area. Change the **X** and **Y** dimensions by clicking on the number to enter **Edit** mode and adding a +1 after the number shown. Leave the **Z** dimension exactly as it is.

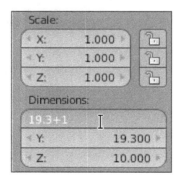

3. Now, select the **Ring** object. Similarly, change the **X** and **Y** dimensions to add 1 to them as well.

   This will make the finger hole 1 mm wider and the ring 1 mm wider as well. This way, the ring itself says the same thickness. If you don't feel as if 1 mm is enough added space, then add more to your ring. You can also make it smaller if it's too loose.

4. Save (*Ctrl + S*) and export the model again. Print it out and test again. Keep iterating and resizing the ring until a comfortable fit is achieved. When you've got it, hide the finger object and move on.

Adding an SD card holder

Now it's time to mount an SD card holder on the ring. The first step is to make a virtual SD card in Blender.

Organizing by layers

When working on a project with multiple parts in Blender, it's possible to organize things by using layers. While the idea of layers may not make as much sense in 3D as in a 2D graphics editor, the idea is similar. Shapes in different layers may share the same space, and they can be looked at and edited separately and independently. Layers are found in the menu at the bottom of the **3D View**, as shown here:

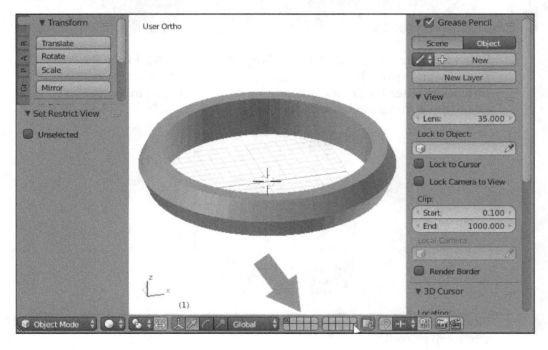

By default, the ring was created in the first layer, and there is a dot in the first layer to indicate that there is an object in it. To switch to an **Empty** layer, simply click on a square on the layer icons at the bottom of the **3D View** or type a number from the top row of the keyboard.

The number keys across the top of the keyboard (unless you're emulating the number pad) from *1* to *0* will switch to the corresponding layer in the top row. Holding *Alt* while pressing a number key will correspond to the rows in the bottom, for a total of 20 layers:

Now the scene looks empty, but checking the outliner panel, you can see that the ring and **Finger** objects clearly aren't gone. To see them again, simply click on the first layer with the object icon, or press the number key *1* at the top of the keyboard.

Now that you know the things you've created aren't gone, go back to the second layer.

Creating a virtual SD card

To create an SD card, perform the following steps:

1. Add (*Shift + A*) a cube to the scene.

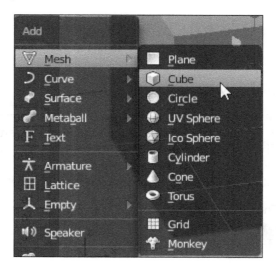

We will be employing a trick to keep the cube above the **XY** plane as we change its size. Instead of moving the cube object up, we'll enter **Edit** mode and move all the points of the cube up. By doing this, the origin of the cube remains where it is and resizing the object will resize it around that origin. This will make more sense when seen practiced.

2. Enter **Edit** mode (*Tab*) and select all (*A*) points.

3. Then, grab (*G*) and move them 1 unit in the z axis (*Z* + *1*). Press *Enter* to end the grab operation:

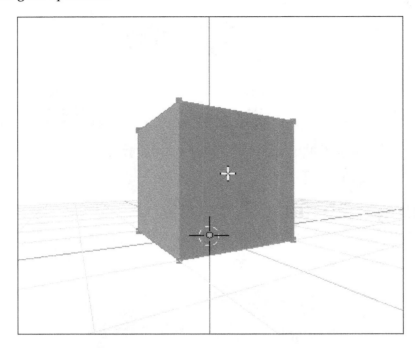

4. Now exit **Edit** mode (*Tab*).

   The cube now sits nicely on the **XY** plane, but more than that, its origin hasn't moved. So now, the bottom face of the cube is resting on the **XY** plane. Experimentally scale (*S*) the cube and move the mouse. Notice that with the cube's origin where it is, the scale operation is only scaling the cube above the **XY** plane. Undo (Ctrl + Z) any experimental scaling before continuing.

5. Now, resize the cube to the size of a SD card. To do this, in the **Properties** panel (*N*), locate the **Dimensions** area and change the dimensions to X: 2.2, Y: 24, and Z: 32:

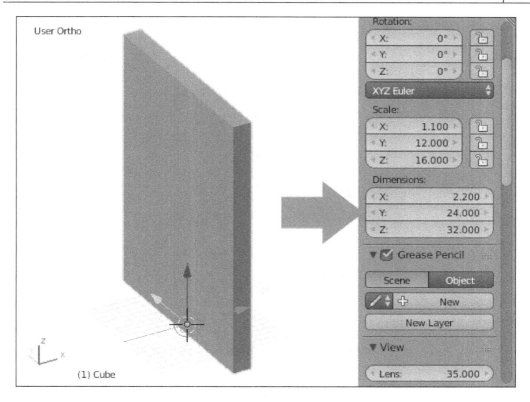

6. In the **Properties** pane on the right-hand side, in the **Object** tab (orange cube), rename this cube to SD Card.

7. Add (*Shift* + *A*) a new cube. Repeat the procedure of entering **Edit** mode (*Tab*) and moving (*G*) its vertices 1 unit along the z axis (*Z*), and then exit **Edit** mode (*Tab*).

8. Now, in the **3D View's Properties** panel, change this cube's dimensions to X: 6.2, Y: 28, and Z: 12.

9. Rename this cube SD Holder:

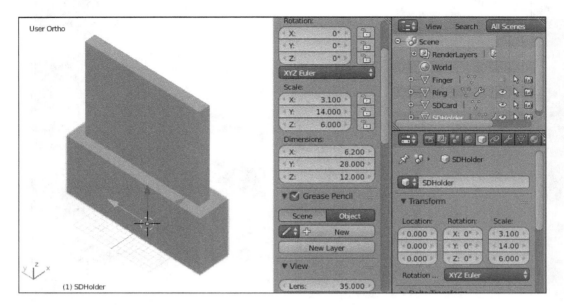

This makes the holder 2 mm thicker than the SD card all around, short enough to easily remove the SD card, while tall enough to hold it securely.

10. Add a **Boolean** modifier to the **SDHolder** object and **Difference SDCard** from it. Hide (*H*) the SD Card object and rotate the view to check the SD Holder object.

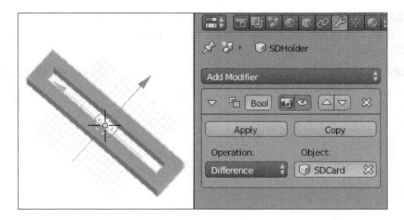

A hole all the way through was not the desired effect. The SD card might fall out! The Boolean operation has not been applied yet, so changes can still be made.

11. Unhide (*Alt + H*) the SDCard object. Select it and move (*G*) it along the z axis (*Z*) 2 units. Then, hide it again and check the SDHolder object.

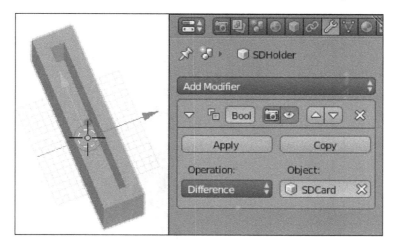

Much better!

Putting it all together

It's now time to bring the SD holder and ring together:

1. To start with, both the **SDHolder** and **Ring** layers need to be visible simultaneously. To make this happen, hold *Shift* while clicking on the other layer or typing the layer number with the number keys at the top of the keyboard:

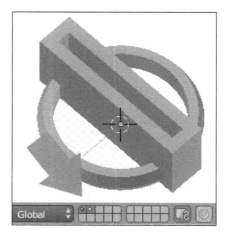

2. Select the **SDHolder** object and move (G) it along the x axis (X) until it is in front of the ring:

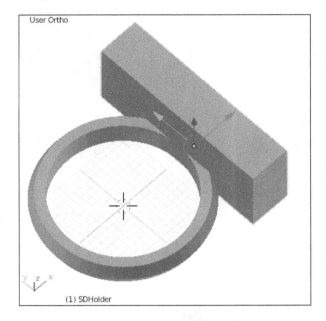

Depending on the view when the last operation was done, a problem may have been observed. What happened to the hole for the SD card? Well, because the **Boolean** modifier was not applied, the hole stayed right where it was, where the hidden **SDCard** object is. If you undo the last movement and redo the movement slowly, you can see the hole stay put as the **SDHolder** object moves.

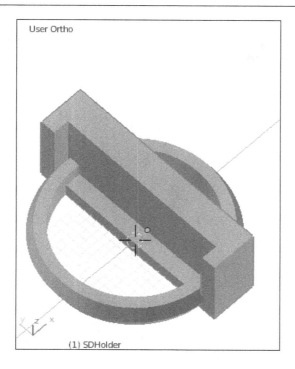

This is amusing but undesired behavior. There are two options to fix it. Either the **Boolean** operation can be applied before moving, or both objects can be moved together. And since there are plans for that hole that include cutting into the ring space, applying the **Boolean** modifier at this time is not the desirable step. So, the hole will have to be moved with the holder.

3. Undo (*Ctrl + Z*) the move operators until the hole is where it belongs in **SDHolder**.

4. Then unhide (*Alt + H*) the **SDCard** object. Select both **SDHolder** and **SDCard** by clicking on one and then holding *Shift* and clicking on the other.

5. Grab (*G*) and move them both in the x axis (*X*) until they are in front of the **Ring** object with ring intersecting **SDHolder**. Make sure there is a good connection here without interfering with the finger hole:

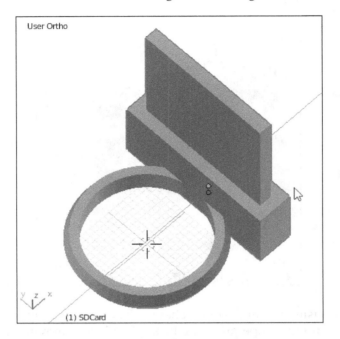

6. Hide the **SDCard** object again, now that everything is in place.

7. Select the **SDHolder** object and add a **Boolean** modifier to Union the **Ring** object to it.

8. Once the **Boolean** modifier has been added, hide (*H*) the **Ring** object to make inspecting the new object easier.

There is a little problem with the **SDHolder** ring right now that is not immediately obvious. However, if seen from certain angles, for instance, looking into the hole from above, it's clear there's a blemish inside the hole. Inside the hole for the **SDCard**, the ring is protruding just a little. If this isn't clear, switch to **Wireframe** view (*Z*) and adjust the view.

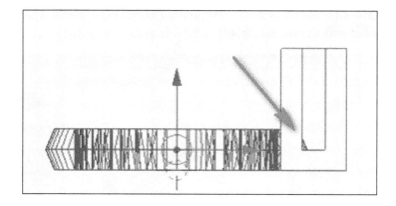

The solution is simple: instead of taking the SD card space out of the holder and then adding the ring to change around the order of the **Boolean** operations, the correct order is a `Union` of the **SDHolder** object with the **Ring** and then the **SDCard** object being differenced from the resultant shape.

9. In the **Outline** view, click on the **SDHolder** object.

10. With **SDHolder** selected, click on the move down arrow button on the first **Boolean** modifier to move it down the stack:

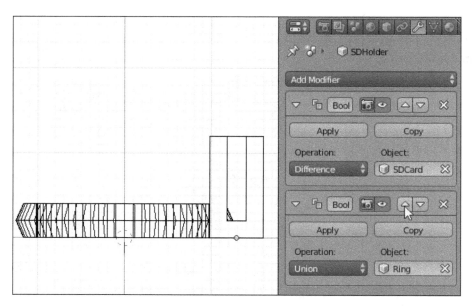

11. The **SDHolder** object is now complete and correct, with the ring attached to the holder and a complete hole for the SD card to sit in:

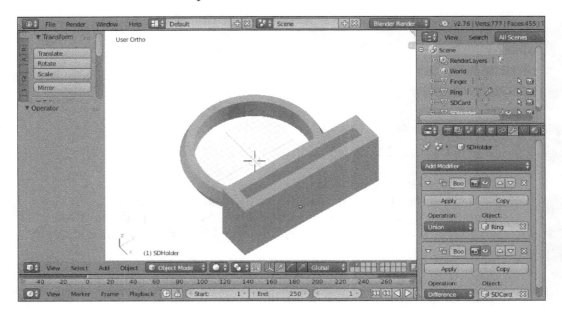

Best of all, by planning carefully, the bottom of the model is flat and ready to print as is. All that is left now is to export (**File | Export | STL**), print out, and use the new SD card holder ring.

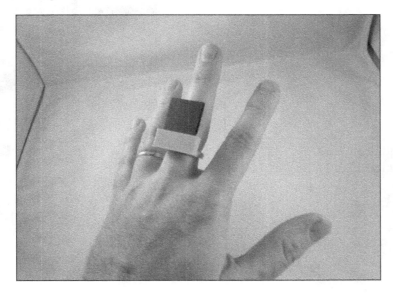

Extra credit

In this project, modifiers were used on simple objects right up to the final object. This allowed the creation of a more complex object while retaining the editability of the simpler objects. However, leaving modifiers unapplied becomes undesirable when objects become too complex. Having to rebuild an object every time an edit is made can crash a computer.

For fun, try to see how quickly Blender will slow down, and maybe try to crash it, by making several cubes and adding a **Subsurf** modifier to increase their polygon count. Then **Boolean** the smoothed cubes together, but don't apply the **Boolean** modifiers. Add more and more **Subsurfed** cubes and add them to the **Boolean** object. Then start moving around and editing the cubes to see when performance becomes choppy. It may be surprising how little it takes, depending on your system:

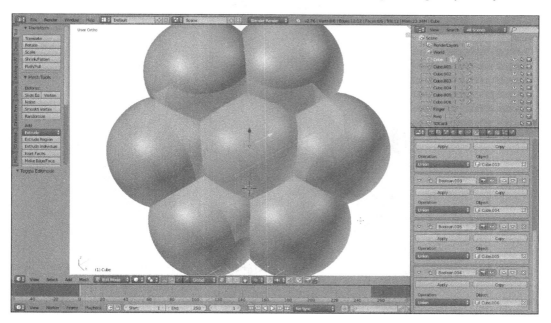

Also, with the measurements of an SD card available, any object can be turned into an SD card holder. Why not make an SD holder key chain or an SD holder that clips to a breast pocket? Add an SD holder to an octopus-shaped pencil holder or any model downloaded online to turn decorative items into something functional. The possibilities are endless!

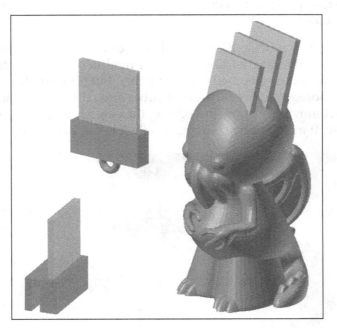

Summary

Modeling things to fit real-life objects is one of the best and most common applications of 3D printing, and Blender is capable of this. Blender may never be able to compete with other CAD programs for precision; nevertheless, with some careful planning and clever manipulation, Blender is capable of extremely precise modeling.

Also, leaving modifiers unapplied is a technique that allows a high degree of flexibility, which is particularly desirable for something that stands a good chance of being customized. If someone else wanted their own SD card holder ring, it would be a simple matter to resize the ring and adjust to position the SD holder.

Now that you have the knowledge and skills to measure and design in Blender, the world can be at your fingertips.

Index

www.ingramcontent.com/pod-product-compliance
Lightning Source LLC
Chambersburg PA
CBHW082123070326
40690CB00049B/4207

* 9 7 8 1 7 8 5 8 8 5 7 3 0 *